PAY NOW OR PAY LATER: CONTROLLING COST OF OWNERSHIP FROM DESIGN THROUGHOUT THE SERVICE LIFE OF PUBLIC BUILDINGS

Building Research Board
Committee on Setting Federal Construction
Standards to Control Building Life-Cycle Costs
National Research Council

NATIONAL ACADEMY PRESS
Washington, D.C. 1991

This report was prepared as part of the technical program of the Federal Construction Council (FCC). The FCC is a continuing activity of the Building Research Board, which is a unit of the Commission on Engineering and Technical Systems of the National Research Council. The purpose of the FCC is to promote cooperation among federal construction agencies and between such agencies and other elements of the building community in addressing technical issues of mutual concern. The FCC program is supported by 14 federal agencies: the Department of the Air Force, the Department of the Army, the Department of Commerce, the Department of Energy, the Department of the Navy, the Department of State, the General Services Administration, the National Aeronautics and Space Administration, the National Endowment for the Arts, the National Science Foundation, the U.S. Postal Service, the U.S. Public Health Service, the Smithsonian Institution, and the Department of Veterans Affairs.

Funding for the FCC program was provided through the following agreements between the indicated federal agency and the National Academy of Sciences: Department of State Contract No. 1030-621218; National Science Foundation Grant No. MSM-8902669, under master agreement 8618641; and U.S. Postal Service grant, unnumbered.

Library of Congress Catalog Card No. 90-64382

International Standard Book Number 0-309-04481-2

Additional copies of this report are available from: National Academy Press 2101 Constitution Avenue, NW Washington, DC 20418

S 331

Printed in the United States of America

BUILDING RESEARCH BOARD (1989–1990)

COMMITTEE ON SETTING FEDERAL CONSTRUCTION STANDARDS TO CONTROL BUILDING LIFE-CYCLE COSTS

Chairman

EDWARD COHEN, Managing Partner, Ammann and Whitney, Consulting Engineers, New York, New York

Members

GREGORY B. COLEMAN, Vice President, American Consulting Engineers Council Research and Management Foundation, Washington, D.C.

NORMAN G. DELBRIDGE, JR., Engineer Consultant, Springfield, Virginia

W. RONALD HUDSON, Professor, Department of Civil Engineering, University of Texas, Austin

ROBERT E. JOHNSON, Associate Professor, College of Architecture and Urban Planning, University of Michigan, Ann Arbor

ALEXANDER I. OUMOV, AIA Germantown, Tennessee

JAMES G. PALMBORG, Operations and Maintenance Division, Embassy Task Group, Sverdrup Corporation, Arlington, Virginia

GARY L. REYNOLDS, Director of Facilities Management, Facilities Planning and Management, Iowa State University, Ames

VICTOR E. SANVIDO, Assistant Professor, Department of Architectural Engineering, Pennsylvania State University, University Park

GAYLAND B. WITHERSPOON, FAIA, Associate Dean, College of Architecture, Clemson University, Clemson, South Carolina

JAMES E. WOODS, William E. Jamerson Professor of Building Construction, College of Architecture and Urban Studies, Virginia Polytechnic Institute and State University, Blacksburg

Friend of the Committee

PAUL V. DOBROW, Value Engineering Manager, Embassy Task Group, Sverdrup Corporation, Arlington, Virginia

Federal Construction Council Liaison Representatives

CAMERON ARNEGARD, HQ USAF/LEEDE, Bolling Air Force Base, Washington, D.C.

ROBERT C. BUMBARY, AIA, Director, Technical Support Services, Office of Facilities, Department of Veterans Affairs, Washington, D.C.

TERREL EMMONS, AIA, Director, Design Support Office, Naval Facilities Engineering Command, Department of the Navy, Alexandria, Virginia

WILLIAM G. ESCHMANN II, PE, Program Manager, Office of Maintenance Management Equipment, U.S. Postal Service, Washington, D.C.

PETER E. GURVIN, Director, Building Design and Engineering Division, U.S. Department of State, Washington, D.C.

WILLIAM H. LEWIS, Program Manager, National Aeronautics and Space Administration, Washington, D.C.

JACK METZLER, General Engineer, Office of Project and Facilities Management, U.S. Department of Energy, Washington, D.C.

GARY RADTKE, PE, Division of Facilities Planning and Construction, Office of Environmental Health and Engineering, Rockville, Maryland

LARRY SCHINDLER, Civil Engineer, HQUSACE (CEMP-EC), Office of the Chief of Engineers, Department of the Army, Washington, D.C.

FRED SUHM, Value Engineering Officer, U.S. Department of State, Washington, D.C.

GORDON WILCOX, PE, Chief, Facilities Management, Aberdeen Area, IHS, Aberdeen, South Dakota

TOM WILLIAMS, Senior Architect, Facilities Standards and Technology Division, General Services Administration, Washington, D.C.

CHARLES YANCEY, Structural Research Engineer, National Institutes of Standards and Technology, Gaithersburg, Maryland

Public Facilities Liaison Representatives

RICHARD W. BLAES, Chief of Maintenance, Department of Facilities and Services, Montgomery County, Rockville, Maryland

ROBERT J. BOEREMA, Director, Division of Building Construction, Department of General Services, State of Florida, Tallahassee, Florida

ROBERT L. WOODWARD, Division of Building Construction, Department of General Services, State of Florida, Tallahassee, Florida
BUCK KATT, Division of Design and Construction, Office of Administration, State of Missouri, Jefferson City, Missouri
KEITH D. KELLY, Department of General Services, State of Maryland, Baltimore, Maryland
RONALD L. THOMPSON, Department of General Services, Division of Engineering and Building, Commonwealth of Virginia, Richmond, Virginia

Advisers to the Committee

ALASTAIR G. LAW, MMP International, Inc.
HAROLD E. MARSHALL, National Institute of Standards and Technology

Project Staff

ANDREW C. LEMER, Director
PETER H. SMEALLIE, Executive Secretary, Public Facilities Council
PATRICIA M. WHOLEY, Staff Associate
JOANN V. CURRY, Senior Secretary

PREFACE

This report is one of several produced by recent studies of committees working—under the auspices of the Building Research Board—on issues related to the use of public assets. Preceding generations have made substantial investments in our public buildings and other facilities that comprise the essential physical infrastructure for our society and, even in times of fiscal stringency, we continue to make new investments. Yet we are coming to recognize that the procedures by which we manage these valuable assets, as well as how we decide where and when to invest, may be serving us poorly. We have seen in recent years a mounting public concern about "America in ruins" and our economy's "fragile foundations," but we have been unable to take effective national action.

The Board believes that more effective management of the public's assets is an area of strategic importance to the nation, an area in which improvements can be made through practical action by committed government officials, professionals, and private citizens working at local, state, and federal levels. We hope that work such as that reported here will help to assure that these improvements are effective.

Andrew C. Lemer, Ph.D.
Director, Building Research Board

EXECUTIVE SUMMARY

A building is an investment made by owners in anticipation of the shelter and services it will provide to the people and activities it will house. With proper management of this investment, returns may continue for hundreds of years, but failure to recognize the continuing costs of ownership can lead to premature loss of services and deterioration of the building and high costs for the building's users. Some materials and building systems are particularly reliable or durable and repay their higher initial costs with savings in future operating and maintenance efforts. Other materials or systems may be selected because their lower initial costs meet the limits of available construction budgets and, with proper use, are likely to deliver entirely satisfactory service. Sometimes safety, security, or aesthetic concerns warrant both higher initial and future costs. Designers and owners of buildings recognize that there are many such choices and trade-offs among initial construction costs, recurring operations and maintenance (O&M) costs, and building performance. Decisions about a building's design, construction, operation, and maintenance can, in principle, be made such that the building performs well over its entire *life cycle* and the total costs incurred over this life-cycle are minimized.

In practice, defining and controlling this life-cycle cost are difficult. The future behavior of materials and mechanical and electrical systems is uncertain, as are the future uses of the building, the environmental conditions to which it may be exposed, and the financial and economic conditions that influence relationships between present and future costs. Unexpected use of the building, unusual events such as storms or earthquakes, poor construction practices, changes of ownership, budgetary constraints, or financial conditions may alter the strategy for minimizing life-cycle cost. Finding the best course of action and assuring that it is followed are challenges that continue as long as a building is in use, challenges that life-cycle cost analysis can help decision makers to meet.

Life-cycle cost analysis is an economic evaluation tool for choosing among alternative building investments and operating strategies by comparing all of the significant differential costs of ownership over a given time period in equivalent economic terms. An effective life-cycle cost analysis depends on having a reasonable range of possible alternatives that are likely to deliver equally satisfactory service to owners and users over a given service life. For projects whose scale does not warrant explicit development of design alternatives, design criteria and guide specifications can help assure that principles of life-cycle cost analysis are reflected in specific designs.

Substantial obstacles to implementing life-cycle cost control in practice include (1) failure of designers to include life-cycle cost goals in their design criteria; (2) failure of owners or managers with short-term responsibility for a building to consider effectively the longer-term impact of their decisions on the building's O&M requirements; (3) general desire of many decision makers to minimize their initial expenditures in order to increase return on investment, meet budgetary restrictions, or both; and (4) lack of data and accepted industry standards for describing the maintenance effect and operational performance of building components. Managers from federal, state, and local government agencies encounter these obstacles in legislative budget procedures; procurement regulations that limit design specificity to enhance competition; and administrative separation of responsibilities for design, construction, and maintenance.

Several decades of experience with highways, and more recently bridges, suggest that improved life-cycle cost management for public buildings can be achieved through development and application of systematically structured and comprehensive life-cycle cost data bases, education and training of professionals and technical staff involved in all stages of the building's life-cycle, and research to develop reliable tools to forecast building performance. A strong and long-term commitment will be required to overcome the obstacles to effective life-cycle cost management of public buildings. In the near term, design criteria may be a practical tool available for controlling *life-cycle costs*, but over the longer term there is a broader range of actions that each agency responsible for these buildings should take:

- Formally recognize control of life-cycle cost as an essential and effective element of the agency's mission.
- Include explicit assessment of design alternatives that influence life-cycle cost as an element of the scope of work and fees of agency designers.
- Assure that value engineering programs and construction contract incentives and other procurement mechanisms demonstrate savings in expected life-cycle cost rather than construction cost only.
- Direct designers to document clearly their design decisions made to control life cycle cost and the subsequently expected operating consequences for each facility.

- Implement cross-training and staff exchange of design and operations and maintenance management personnel to assure that life-cycle cost management principles and understanding of how life-cycle cost is controlled at all stages in the facility's service life are applied in practice.
- Establish a life-cycle cost management system to maintain O&M data and design decisions in a form that supports operations and maintenance management and feedback of O&M experience to future facility designs.
- Assign accountability for maintenance and repair at the highest levels in the agency. Responsibilities should include effective use of maintenance and repair funds and other actions required to validate prior decisions on facility life-cycle cost. Assure that adequate resources are available to implement life-cycle cost management decisions.

CONTENTS

PAY NOW OR LATER: CONTROLLING COST OF OWNERSHIP FROM DESIGN THROUGHOUT THE SERVICE LIFE OF PUBLIC BUILDINGS

1

INTRODUCTION

Public buildings are assets needed to serve government purposes. The public is called upon to invest in these assets and pay the costs of their upkeep. Minimizing the total *costs of ownership*[1] is the most efficient use of the public's resources to obtain the services these assets provide. Overcoming the economic, technical, and political obstacles to meaningful control of the total costs of public buildings will enhance productivity and the public's return on its investment.

Agencies' managers generally recognize the need for facilities that serve efficiently the purposes of government and seek ways to overcome obstacles to effective cost management. The sponsors of the Federal Construction Council (FCC)[2] asked the Building Research Board (BRB) to undertake a study of the use of life-cycle cost analysis—a formal set of principles and procedures for considering total costs of ownership—for setting building design criteria. This document is the final report of that study.

[1] Italicized terms are defined in Appendix A.

[2] This study was requested and sponsored by the FCC, a group of 16 federal agencies with responsibilities and interests in construction and building research. These agencies have annual construction budgets totaling more than $7 billion.

The BRB established a committee[3] of experts with broad expertise and experience to undertake this study and to recommend how agencies could improve their current practices. Meeting over the course of approximately 1 year, committee members discussed a variety of complex technical and management issues, drawing on their experience to assess the current state of knowledge and make the recommendations presented here.

The committee's work was motivated by federal agencies, but the topic of study is relevant to architects and engineers, owners, and managers of all government buildings and other constructed facilities. To the extent that facilities are built and used by the same institution, the same concerns apply in the private sector as well. Buildings are an investment in the future, and substantial expenditures of funds for design and construction are made by a building's owner in anticipation of the shelter and services the building will provide to the people and activities it will house. Structures around the world demonstrate that the returns on such investment may continue for hundreds of years.

Some public facilities provide centuries of service.
West Front of the U.S. Capitol (Photo courtesy of Amman & Whitney)

[3] Biographical sketches of the committee's members are presented in Appendix B.

These returns are seldom achieved without continuing effort. Owners must make expenditures for labor and materials to operate and maintain a building, expenditures that continue until the building is demolished or abandoned. Structures around the world also demonstrate that failure to make these expenditures effectively can lead to premature deterioration or loss of services and damage to the facility, expose occupants to unsafe and unhealthy conditions, and impose additional costs on the building's users.

Some materials and building systems are particularly durable and repay their higher initial costs with savings in future operating and maintenance (O&M) expenditures. Other materials or systems may be selected because their lower initial costs meet the limits of available construction budgets and, with proper use, are likely to deliver satisfactory service. Some design choices raise the cost of construction, operations, or both but also increase the service productivity or revenue received from the completed building.

Designers and owners of buildings often recognize that there are such choices and trade-offs between initial construction costs and recurring O&M costs and that decisions about a building's design, construction, operation, and maintenance can be made—in principle—so that the building performs well over a specified period of time and the total of all costs incurred over that period will be minimized. [4]

In practice, defining the design options and operating strategies that will lead to the lowest *life-cycle cost* is difficult and subject to uncertainties. The behavior of materials, and mechanical and electrical systems must be forecasted, along with the likely uses of the building and the environmental conditions to which it may be exposed. Financial and economic assumptions and the period of time over which the analysis is made will influence the results. Analysts have devised a variety of ways to deal with these uncertainties.

However, a variety of factors may subvert effective action. Sometimes budget constraints impose pressures to reduce construction costs and lead in turn to design choices that raise O&M requirements. Similar pressures in the planning and design stages may underlie neglect to perform analyses and reduced effort to develop feasible alternatives that would save money in the long run. Sometimes O&M efforts may not achieve results envisioned in design because of later budgetary pressures, lack of staff understanding of the designer's intent, poor information, or human error. Fires, earthquakes,

[4] This report is oriented primarily toward public facilities and gives only limited attention to the revenue-producing possibilities of facilities. The committee assumed that net revenues in excess of costs—that is, positive returns on the government's investment in facilities—are best attributed to the functional program or mission the facility supports. That is, the government does not undertake real estate development or facilities management as a business proposition, although the committee noted cases where agencies (e.g., the Postal Service and General Services Administration) are being encouraged to act as commercial developers.

violent storms, or other unusual events may damage facilities. Unanticipated use of the building, changes of ownership, or financial conditions may alter the strategy that would minimize life-cycle cost.

Because of these factors, facilities designers, builders, owners, and managers must continue working to control total costs of ownership throughout a facility's *service life*. Life-cycle cost analysis is typically used in planning and design, but the committee found that other opportunities for using life-cycle cost analysis lie beyond these early stages.

The committee found, however, that the most difficult obstacles to controlling total costs of ownership are those raised by administrative procedures and managerial or political decisions driven by short-term gains. Budgeting processes that divorce capital and operating expenditures make it difficult to identify and manage total costs of ownership. Their limited tenure may encourage senior managers and elected officials to value immediate results over long-run efficiencies. Competing public demands for government action may push these officials to shift resources away from facilities needs and toward those issues that attract strong constituencies. These obstacles impose cost burdens on the public and must be overcome if the greatest return on the public's assets is to be achieved.

The following chapters present the committee's assessment of the problems that federal agencies and others encounter in trying to construct buildings that yield effective service at low ownership costs and how life-cycle cost analysis may be used to help solve these problems. Chapter 2 briefly describes the underlying principles of life-cycle cost analysis as they relate to the real problems faced by facilities, designers and managers. Chapter 3 considers the substantial obstacles to effective application of the principles of life-cycle cost management in the institutional environment within which government agencies must operate. Chapter 4 reviews how government agencies are attempting to achieve life-cycle cost management and how lessons learned in highway pavement and bridge management could improve control of life-cycle cost of buildings. Chapter 5 presents the committee's recommendations for long-term and immediately implementable actions that federal agencies can take to achieve better control of life-cycle costs of their facilities.

Several appendixes present additional background information: a glossary of terms (Appendix A), a brief review of the economic principles and procedures of life-cycle cost analysis (Appendix C), and a preliminary listing of the types of questions and decisions that a building life-cycle cost management system should be designed to address (Appendix D).

This report is not intended to be a comprehensive discussion of the principles and state of the art of life-cycle cost analysis but rather a presentation of the reasoned views of a committee asked to give advice about a technical matter within the decidedly nontechnical context of how government facilities are planned, designed, constructed, operated, and, most importantly, used by people serving the ends of government.

2

LIFE-CYCLE COST AND ITS CONTROL[5]

A building or other constructed facility provides shelter and service to its owners and occupants over a period of years. As this period extends to decades, portions of a building may be altered, extensively repaired, or replaced. Sometimes social, technical, or economic changes render buildings effectively obsolete, and they are abandoned, demolished, or replaced entirely. Sometimes buildings achieve particular artistic, cultural, or historic distinction that leads to their preservation when they might otherwise be retired from service.

While the number of years that a building or its component parts will be used is uncertain, building owners and designers nevertheless think in terms of a distinct *economic life* when making decisions about what materials and equipment to use in a new building and how much to spend on construction and the other costs of ownership. Construction costs are only a small portion of the costs of ownership, and the building owner who recognizes that he or she will bear not only these initial expenses but also the future costs of the building's operation, maintenance, and use should have some interest in controlling all of these costs. Determining how to assess, compare, select, and then control these costs so that a building will provide adequate service throughout its life is the subject of life-cycle cost analysis.

[5] Appendix C briefly discusses the major principles underlying life-cycle cost analysis.

WHY CONSIDER LIFE-CYCLE COST?

Life-cycle cost analysis is an economic evaluation process developed to assist in defining and then deciding among alternative building investments or operating strategies. Alternatives are characterized by differing patterns of costs likely to be incurred throughout a given time period, and the analysis seeks to assess these different patterns in comparable terms. Often, a single number—the *present value* of life-cycle cost, for example—is used to indicate a preferred choice among alternatives.

Alternatives are typically defined to illustrate in a systematic way some trade-off between first costs (e.g., for construction or equipment procurement) and future *recurring costs* (e.g., for maintenance or electric power consumption). The analysis is often undertaken with an expectation that an alternative can be found that will have the lowest life-cycle cost (see Figure 2-1).

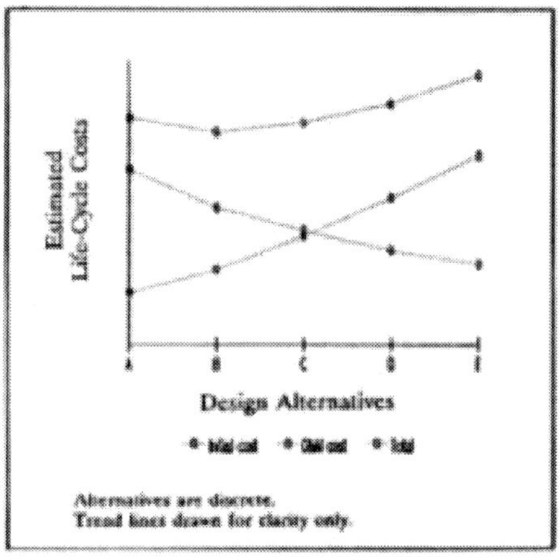

Figure 2-1 Trade-offs to minimize life-cycle cost.

Finding the minimum life-cycle cost alternative generally requires good understanding of the technical factors underlying the tradeoff being considered, reasonable estimates of the various costs involved, and a certain degree of ingenuity and judgment. *Designers* must be familiar with how choices of materials, facility configurations, and other design variables influence initial cost and future operations and maintenance. The principles underlying life-cycle cost analysis imply that there is a continuum of alternatives, but in fact building-related options are generally quite distinct and discontinuous (e.g., only certain distinct actions and combinations are practical). The designer or manager must exercise some imagination to define a range of alternatives that illustrate realistically the trade-offs to be made and that indicate a minimum life-cycle cost.

Sometimes alternatives are simply variations of a particular design or operating strategy. Each change that yields a further reduction is in effect another alternative that is compared, in principle, to the initial design, the "base case." The analysis stops when further significant reductions cannot be found.[6] (See box next page.) Care is required to assure that performance is not compromised in the search.

This search for the minimum life-cycle cost—and the resulting efficiencies of resource utilization that a search implies—makes life-cycle cost analysis particularly appropriate for use by government agencies. These agencies build and use facilities to serve their missions, and the costs are paid with public funds. Even when one agency is responsible only for construction of a facility to be used by another agency, they all share a responsibility to use public funds as effectively as possible, now and in the future. The public should not care, in principle, whether funds are spent for construction, operations, or maintenance, today or tomorrow, so long as no more is spent than is necessary. In practice, as will be discussed in Chapters 3 and 4, the public, its elected officials, and its career civil servants may care a great deal about when and for what purposes funds are expended, and their concerns can be major obstacles to life-cycle cost control. The committee recommends that government agencies increase their use of life-cycle cost analysis as a means to overcome these obstacles.

IMPORTANT ASSUMPTIONS

The principles of life-cycle cost analysis include several assumptions that can have important consequences for design and management, particularly when the assumptions are implicit and may not be recognized by the user of the analysis results.[7] Foremost among these assumptions is that all costs are measurable in monetary terms. The life-cycle cost estimate is only one of several factors that may be important to selecting among several designs or operating strategies, and the analysis is often restricted to *financial costs* alone—actual transfers of funds to purchase building-related goods and services.[8]

---

[6] In practice, each variation may be compared only to its predecessor, in search of marginal improvements. If functional performance is explicitly considered, this process may be termed *value engineering*, but some value engineering studies (particularly those associated with construction contractor incentive clauses) address construction costs only. The term life-cycle value engineering is sometimes used to distinguish studies that do address life-cycle costs from those that focus on construction cost alone.

[7] Some of the computational assumptions made in analysis can significantly influence the results. These assumptions are discussed in Appendix C.

[8] Economists have developed sophisticated procedures for evaluating in monetary terms a wide range of benefits and costs, but these methods are not typically used for life-cycle cost analysis and were beyond the scope of the committee's deliberations.

VA Analysis of Potential Life-Cycle Savings

The Department of Veterans Affairs (VA), with more than 172 medical centers (some 2,000 buildings), owns and operates the nation's largest hospital system. The VA has developed a hospital building system designed to be economical while easing adoption of changing health care needs, simplifying bidding, and promoting fire and earthquake safety. The first demonstration of this building system was the VA hospital in Loma Linda, Calif., completed in 1973.

A major force in the system's development was the effort to lower the VA's total long-term costs as owner of the facility. Detailed analyses were made of construction, operating, maintenance, and renovation costs anticipated over the 40-year economic service life typically assumed as the basis for the agency's planning and design decisions. This analysis highlighted seven major factors that could reduce the costs of ownership (the dollar estimates, based on prevailing conditions in the early 1970s, would probably be higher today):

1. Elapsed time to plan and build the facility. The new system seemed likely to cut 6 months or more from the typical 63-month project time from start of master planning to end of construction. The capital cost savings were estimated then to exceed $350,000 on a 500,000-square ft hospital.
2. Building size, configuration, and complexity. Standardization of building components and concentration of utility systems offered potential construction cost savings of 3 percent, compared to conventional designs.
3. Space utilization. Careful programming and improved layout of mechanical systems offered a reduction of 4 percent in gross built area needed to meet a given set of requirements, with consequent savings in maintenance, operations, and housekeeping costs as well as construction costs.
4. Maintenance and minor alterations. Better materials performance and better access to service elements were estimated to offer potential annual cost reductions of 10 percent, compared to conventional designs.
5. Utility usage. Optimization of the utility systems, especially the HVAC (heating, ventilation, and air conditioning), were estimated to yield savings of 5 percent in annual costs.
6. Housekeeping. Incorporating such design features as adjustable-height loading docks, wall-hung plumbing features (for easier cleaning), and smooth walls and doors (for easier washing) offered annual savings estimated to amount to 5 percent.
7. Major rehabilitation. The system's design will facilitate major rehabilitation, offering cost reductions estimated to be at least 30 percent compared to conventional designs.

The present value of the resulting combined savings over a 40-year period, calculated using a 5 percent discount rate, was estimated to be approximately 10 percent of the life-cycle costs of ownership, with nearly three-quarters of the savings due to reductions in O&M costs. The total life-cycle costs were estimated to be approximately 8.7 times the initial costs in this calculation, so the relative saving would be less if a higher discount rate had been used.

(Source: Feasibility Study--VA Hospital Building System, Research Study Report, Project No. 99-R003, prepared by the Joint Venture of Building Systems Development and Stone, Marraccini and Patterson for the Research Staff, Office of Construction, Veterans Administration, U.S. Government Printing Office, Washington, D.C., 1968)

A second key assumption is that all alternatives considered in the analysis deliver the same performance throughout their service lives. This assumption is violated, for example, when decisions made to reduce life-cycle energy costs have resulted in indoor air quality problems. The committee noted that a growing body of evidence suggests that apparent reductions in life-cycle costs of air conditioning and lighting systems may be far exceeded by costs of lost productivity of the work force housed in a building when performance is not maintained (Woods, 1989; Brill, 1984). Analysts suggest that lost work time, insurance costs, lost revenue, and other financial consequences of poor building performance should be explicitly considered in the life-cycle cost analysis, but doing so has not become common practice. [9]

"Interstitial space" in the VA hospital building system facilitate updating of mechanical system, extending the structure's potential service life. (Photo courtesy of Department of Veterans Affairs)

Finally, life-cycle cost analysis assumes there is an easy interchangeability of present and future costs, for example, spending more initially to purchase

[9] Committee members noted that federal government agencies—in principle responsible for wise application of taxpayers' funds without regard to whether construction, operations and maintenance, or occupants' costs are at issue—would do well to explore this extension of life-cycle cost analysis.

durable materials or systems to achieve future savings in a building's maintenance costs, or choosing less costly and durable materials and systems that can be maintained at higher cost (i.e., by painting or lubrication) to yield the same service. The analysis may fail to deal explicitly with probabilities that a system or material may fail prematurely, that maintenance efforts may be ineffective, or other factors that represent uncertainties in cost estimates. These uncertainties increase as one tries to project further into the future.[10] Also, as will be discussed in Chapter 3, forces at work in the public sector may tend generally to invalidate this assumption.

LIFE-CYCLE COST MANAGEMENT

Life-cycle cost analysis is used to consider alternative future courses of action: designs for facilities yet to be built; strategies for future operations; maintenance, repair, and renewal programs. Selection of a preferred alternative implies that certain actions will be taken in the future. Effective life-cycle cost management can be achieved only if these actions are in fact taken, and failure to do so may raise the costs of ownership well above the levels anticipated in the analysis. Spending below targets set for normal maintenance, for example, may substantially increase costs of repair, replacements, and loss of use, costs that might have been avoided. (See box.) In general, inadequate operations and maintenance efforts will raise a facility's costs of ownership.

Hoping to delay normal wear and aging of a facility, managers may set future spending to exceed targets defined in the life-cycle cost analysis. However, unless the initial analysis is somehow faulty, savings resulting from extended life of building components or avoided damages are unlikely to balance fully the increases in maintenance spending. The total costs of ownership will again be raised above anticipated levels. In some cases higher life-cycle costs are warranted by the unique strategic, historical, or symbolic character of the facility. Buildings such as the U.S. Capitol, the White House, state capitol buildings, and city halls, as well as a wide variety of other landmarks, will be preserved and renovated regardless of the costs of dealing with difficult spaces and obsolete materials and operating subsystems. Costs of ownership may be

[10] Theorctical analyses suggest that policy makers and the public at large, recognizing the risks that future spending will not occur as planned, will generally prefer to build facilities with longer service lives (Glazer, 1989).

Costly Consequences of Neglect

Subject to the dictates of Congress and the demands of their tenants, the General Services Administration (GSA) owns and manages more than 7,000 federal buildings. This inventory includes the Pentagon, still the world's largest office building, with some 6 million gross square feet of space and five decades of service.

The mechanical equipment illustrates the broader problems that leave some members of Congress, quoted in the Washington Post, "appalled at the severe deterioration of the Pentagon Reservation": five coal-fired boilers linked to 1.5 miles of steam tunnels and a series of chillers drawing water from the Potomac River were installed when the Pentagon was constructed in the early 1940s. The GSA contracted in 1973 to replace the furnaces with oil-fired boilers, and the first unit was nearly completed when the oil embargo started. Replacements were canceled, and the coal-fired boilers continued to be used. For the following 15 years, shortages of funds, disputes over responsibility for maintenance, and inability to gain congressional or local government permission to proceed with plans hampered action to rehabilitate the system. In 1987 boilers were rented--at an annual cost of $1.2 million--to do the work of the equipment that is so decrepit, Pentagon officials told the Post, that it might fail or explode at any minute.

The boilers are but one symptom of the Pentagon's broader deterioration that led one senator to term the building, according to the Post, "a wreck. . . about to fall in." The Defense Department's proposal to spend $1.1 billion on repairs and office upgrading encountered serious opposition in the Senate appropriations military construction subcommittee.

While such comparisons have limited validity, the 1988 simple average construction cost of approximately $88 per square foot for new federal office buildings (excluding land expenses) suggests how high this price tag may be. GSA's lack of money and political strength within the administration, continuing turnover of senior management personnel (including 17 administrators since 1972, only six of whom won Senate confirmation), and weak management support systems are blamed by some observers. Others cite Congress's or the White House's indifference to property management and maintenance and the significance of GSA's role. Whatever the cause and cost, congressional conferees have come to recognize that "failure to maintain the Pentagon means that a massive renovation . . . is now needed to bring the Pentagon up to a satisfactory level."

(Source: Judith Havemann, "Pentagon's Maintenance Woes: A Boiling Issue for GSA," Washington Post, and John Morawitz, "Analysis of Available Statistics on Comparative Construction Costs," Technical Report No. 94, Comparing the Construction Costs of Federal and Nonfederal Facilities (Summary of a Symposium), Federal Construction Council, National Academy Press, Washington, D.C., 1990).

minimized within the context of preservation but may be higher than those for functionally comparable new construction.[11]

[11] On the other hand, the costs of replicating in new construction the detailing, craftsmanship, and performance characteristics of many historical structures would be outrageously high. While private sector owners are able

RELATIONSHIP OF PLANNING AND DESIGN CRITERIA AND LIFE-CYCLE COST

An owner or his or her agents in procuring new construction typically establishes a set of criteria that describe the characteristics the new building should have to deliver adequate performance over the course of its future *lifetime*. Many purchasers implicitly accept some criteria established by others, as is the case when private construction must meet the criteria stated in local building codes. [12] Architects and engineers effectively establish some of the criteria for their clients by following standard design practices or adopting specifications developed by manufacturers of materials and equipment. Some large corporations, government agencies, or other organizations that procure and manage large *portfolios* of buildings develop their own comprehensive statements of all of the criteria they expect designers to apply in a new building's design, but even in these cases many of the criteria are adopted from elsewhere. These criteria concentrate most on design and construction of the facility and often do not address operations and maintenance (O&M).

Regardless of their source, these criteria influence a new building's cost in complex and sometimes unforeseen ways. The number and complexity of interactions of these many criteria[13] preclude generalized analysis, and it is in designing a specific building that the architect or engineer may find that local conditions, building program, and design criteria combine to require or preclude particular materials, equipment, or design details. Nevertheless, there is sufficient regularity in the relationship of criteria, designs, and costs that professionals can try to establish their general criteria such that the final design of each specific building possesses the characteristics desired.

Recognizing that certain materials or equipment cost more initially but offer O&M savings and lower life-cycle cost than other choices, owners may establish design criteria that require these materials or equipment to be used. For private sector owners and designers, these criteria may be as simple and straightforward as the specification of a particular brand-name product. However, in the interest of encouraging free competition, federal government agencies must typically avoid such direct statements and instead develop

---

to recapture cost increases in the higher rents that historical or architecturally unique buildings may command, public agencies must rely on the sensitivity of the public at large and elected officials to recognize the value purchased with higher life-cycle cost

[12] A relatively unsophisticated purchaser may give little explicit attention to these criteria, depending instead on building codes and standard design practices.

[13] The manuals and guidelines documents that some large corporations and federal agencies expect their designers to follow fill thousands of pages and many volumes.

generic criteria describing the performance required of building components. Agency professionals thus are challenged with developing generic criteria that will minimize life-cycle cost and remaining robust in the face of efforts to reduce the costs of building construction.

REFERENCES

Brill, M., 1984, Using Office Design to Increase Productivity, :Work Place Design and Productivity, Inc., Buffalo, N.Y., pp. 337–352.

Glazer, A., 1989, "Politics and the choice of durability," The American Economic Review, December, pp. 1207–1213.

Woods, J. E., 1989, "Cost avoidance and productivity in owning and operating buildings," Occupational Medicine: State of the Art Reviews, vol. 4, no. 4, pp. 753–769.

3

OBSTACLES TO LIFE-CYCLE COST CONTROL

The idea that life-cycle costs can be controlled and minimized has wide appeal, but life-cycle cost analysis has not been consistently applied in the design and management of buildings. A variety of factors associated with the methodology limit acceptance and practical application of the principles and procedures of analysis.

DATA AND PROCEDURAL OBSTACLES

The analysis procedures themselves present a number of problems that discourage use of life-cycle cost. One of the most difficult problems is the shortage of reliable information on historical costs and performance, which is needed for accurate estimation of costs.

First, buildings are dissimilar, located in different areas, built at different times, and operated by a variety of owners and their agents. Cost data are therefore difficult to collect and analyze. Second, there is no institutional mechanism—beyond the federal government and federal buildings—for pulling together data from many sources.[14] Third, accounting systems used by

[14] Organizations such as the Building Owners and Managers Association and the American Society for Heating, Refrigerating, and Air-Conditioning Engineers assemble data on aggregate operating costs of major types of buildings or major building subsystems. The U.S. Army Corps of Engineers Construction Engineering Research Laboratory's past efforts to assemble such

building managers and contractors seldom make it possible to identify accurately the costs of maintenance and repair of specific components (e.g., the roof) of single buildings.

Another problem with analysis procedures, closely associated with lack of information is the uncertainty inherent in any forecast of future costs and performance. How a building and its functional subsystems behave—over the course of a 25-to 30-year service life—depends on the building's design and construction, the weather, the buildings' users and operators, and variations in materials. Predictions can be made only in probabilistic terms, and the chances of being right are greater when data upon which to base the predictions are adequate, valid, and timely. If data are lacking and the chances of being wrong are consequently large, life-cycle cost estimates cannot contribute most, effectively to improving the chances of making good design and operations decisions.

A third problem is that the assumptions reflected in selection of parameters used in the life-cycle cost analysis—such as economic life *and discount rate*—may or may not reflect well the conditions that actually occur in the future. The results of the life-cycle cost analysis depend on these assumptions, and different assumptions could indicate that a different course of action should be taken. (See box.)

OUTGUESSING THE FUTURE

"Sensitivity analysis" is the term used to refer to asking "what if . . ." questions about life-cycle costs. "What if electric power costs double; would we still be better off using heat pumps?" "If the roof that uses a low-maintenance but somewhat more expensive new material must be replaced after only 10 years rather than 15, is the extra construction cost still justified?"

Mathematicians, economists, and other speak of a "robust" solution as one that remains the best answer when the assumptions change. Life-cycle cost analysis should be used to help develop robust designs and operating strategies—ones that are likely to be an efficient use of the owner's limited resources for any reasonably likely future conditions.

Finally, the level of effort required in analysis increases rapidly as the number and range of alternatives increase. The ability of an analyst to find a lower life-cycle cost alternative is limited by the funds and time available for analysis.

INSTITUTIONAL OBSTACLES

The institutional context of building design and operating decision making raises substantial obstacles to the use of life-cycle cost analysis to successfully manage the costs of ownership. Some of these obstacles arise from the unique

data underlie life-cycle cost analysis models now being developed.

conditions of decision making in the public sector, while others are common to both private and public sectors.

First, changes in a building's use, commercialization of new technology, and external economic forces or government policies may alter radically the future conditions that a building faces. In such cases a revised life-cycle cost analysis might indicate that different design or operating decisions would have been preferred. Sometimes a building may be sold or demolished and replaced with a more appropriate facility, and it may seem that the effort required to conduct life-cycle cost analysis was wasted. Some critics use such reasoning to argue that there is not enough time, money, or trained people to perform life-cycle cost analysis on a routine basis.

When funds are available only for certain types of cost and not for others (e.g., for new construction but not for maintenance), or when government policies intended to achieve ends unrelated to the facility encourage the use of certain technologies or design options, an agency's decisions may not be made to minimize life-cycle costs of ownership. For example, a local or state government using funding available from a higher-level government agency to construct a public facility may seek to maximize construction costs in order to save on its own future maintenance expenses. Programs to foster energy savings or enhancement of educational facilities may encourage replacement or renovation of heating systems or school buildings sooner than would otherwise have occurred. In these cases the energy efficiency or modern schools may be purchased at higher total life-cycle costs than would otherwise have been necessary.

Most facilities have long service lives compared to the lifetime of agency missions and legislated programs. This makes it inevitable that these facilities will encounter changes in use or introduction of new technologies. When changes occur, investments made in a building to save on future costs may appear, in hindsight, to have been wasted. This appearance then is seen by some as a basic flaw in life-cycle cost analysis. The problem may be more acutely felt in the public sector because government cannot easily realize the *benefits* of historic value, prime location, or architectural merit that may bring higher rents and offset costs in the private sector.[15]

The failure of many jurisdictions to link capital and operating expenditures within a multiyear budgeting framework also poses an obstacle to life-cycle cost control. Legislative bodies sometimes seem unaware (or disinclined to acknowledge) that authorization to construct a facility commits the jurisdiction to many years of operations and maintenance (O&M) expenditures and that reductions in initial costs may raise future expenses. Total life-cycle costs may then be higher than might have been necessary because no provision is

[15] The National Aeronautics and Space Administration faces an extreme example of this problem when it must construct complex and specialized facilities to house new space missions. The agency sometimes makes design choices aimed specifically at achieving a limited service life, matched to its current mission.

made in the budgeting and appropriations process for these future expenses or for initial spending to minimize them.

A more general obstacle is the desire of many decision makers to minimize their initial investment in a building. This desire may spring from an effort to increase return on investment, to meet budgetary restrictions, or both. Because the relationships of design choices and O&M costs are poorly documented, and also because O&M costs are problems for the future, it is often difficult for designers to argue persuasively in favor of designs that will raise today's construction costs to achieve lower O&M costs tomorrow, even if total life-cycle costs can be lowered.

Durability and low maintenance costs have characterized Washington's Dulles International Airport. (Photo courtesy of Ammann & Whitney)

Federal agencies may encounter this obstacle with particular frequency because of the way in which *value engineering* is practiced and applied in construction contracts. Contractors are often encouraged to review a building's design and propose ways to reduce costs with no loss of conformance to agency design criteria. The successful contractor is compensated for his or her effort by a sharing of apparent savings. [16]

[16] For example, the contractor may receive 30 percent of the reduction in the contract cost attributed to the value engineering changes.

Contractors typically have no interest in a building's O&M activities and little concern for their future costs. It is the agency's and designer's responsibility to demonstrate that apparent savings in construction cost will be more than offset by increased O&M effort. When agencies are faced with severe budgetary pressures, it becomes increasingly difficult to hold firmly to design decisions that are not supported by solid technical data and guaranteed future savings.

MANAGEMENT OBSTACLES

Once a particular design or operating strategy has been selected—with its anticipated life-cycle costs—there are management obstacles that may lead to growth of ownership costs during the facility's service life. Some of these obstacles arise from a failure to maintain management commitment to the course of action implied in the initial strategic decision. (See box next page.)

Probably the most serious of these obstacles is the tendency toward deferral of maintenance efforts, which leads to premature deterioration and failure of building components, accelerating increases in costs for repair or renewal, and potential threats to safety and health. Because the relationships of maintenance effort to building performance and other costs of ownership are often difficult to demonstrate, professionals responsible for O&M activity often find their budgets to be below what they feel are adequate levels.[17] Further, government agencies typically maintain a strict separation of responsibilities and budgets for construction versus operations and maintenance, so that managers responsible for one activity have little incentive to use their own resources to achieve savings in the other area.[18]

Another serious obstacle is the lack of accepted industry standards for describing operational performance of all building components. The committee noted that there is relatively little feedback of information from buildings in service to new designs, which might yield a reliable basis for estimating how maintenance effort influences service life of many building components. The consensus-of-experience basis of the great majority of design criteria used in building does not typically consider life-cycle performance, and one review of the situation identified only a single criteria-developing organization that does

---

[17] Another committee of the Building Research Board (BRB) asserted that, in the absence of better information, annual facilities maintenance expenditures should be budgeted at 2 to 4 percent of current facilities replacement cost to avoid growth of a backlog of maintenance requirements (BRB, 1990).

[18] Several members of the committee observed that federal agencies may be better than most state or municipal government agencies in overcoming this obstacle but that it does exist at the federal level.

Politics, Management Psychology, and Failure of Life-Cycle Cost

Practical achievement of low life-cycle cost by balancing expenditures for construction, operation, and maintenance often fails in the face of political reality and the motivations of those who control the budgets.

"If it ain't broke, don't fix it" is one management principle that leads to the failure. New York City "saved" millions of dollars over a period of years by neglecting maintenance of its highway bridges, until resulting structural deterioration forced closure of the Williamsburg Bridge and disrupted thousands of commuters and businesses. Part of the reason for neglect was that the city pays most of the cost of maintenance while state and federal funds were available for a large proportion of major new construction and reconstruction.

Another source of failure to achieve low life-cycle cost is the argument that it makes sense to neglect regular maintenance, accept the deterioration of the facility's performance, and in a few years rehabilitate to fix the problems. This argument neglects the adverse influence that deteriorating facilities can have on the productivity of those who use the facilities. Data to demonstrate this influence are sparse but striking: Economists studying irrigation in Pakistan found that moderate increases in maintenance spending (about 10 percent) had substantial net benefit in increased agricultural production in the irrigated areas. Others studying the consequences of "building-related illnesses" (e.g., respiratory and stress-related problems) estimate that billions of dollars may be the lost in absenteeism and lost productivity by U.S. businesses due to poor maintenance or efforts to save energy costs through reduced performance of heating and air conditioning systems.

Finally, there is the inevitable preference of those in authority for new construction over maintenance, a preference that has some sound theoretical bases. Given a choice between a durable, long-lived project and one designed for earlier rehabilitation, economic analysis shows that voters, politicians, and senior managers appreciate that the durable investment commits the government more firmly to a course of action that the proponents find desirable. Planning for higher maintenance effort or early rehabilitation gives opportunities for future voters or managers to alter this commitment, while the added expense of extending the commitment through marginally greater first costs is relatively inconsequential, once the basic decision to build has been made.

(Source: M. A. Chaudhry and M. Ali, "Economic Returns to Operation and Maintenance Expenditure in Different Components of the Irrigation System in Pakistan," ODI/IIMI Irrigation Management Network Paper 89/1d, Overseas Development Institute, London, June 1989; J. E. Woods, "Cost Avoidance and Productivity in Owning and Operating Buildings," in Occupational Medicine: State of the Art Reviews, Vol. 4, No. 4, Hanley & Belfus, Philadelphia, Oct.-Dec. 1989; A. Glazer, "Politics and the Choice of Durability," The American Economic Review, Vol. 79, No. 5, December 1989).

so.[19] Designers and professionals responsible for adopting design criteria thus must effectively break new ground to set generic criteria to foster lower life-cycle costs.

[19] The American Society for Heating, Refrigerating, and Air Conditioning Engineers Standard 90.2, "Energy Efficient Design of New Low-Rise Residential Buildings," specifies insulation values for the envelope of a building. Development of this standard required several years (Underwood, 1988).

REFERENCES

BRB (Building Research Board), 1990, Committing to the Costs of Ownership: Maintenance and Repair of Public Buildings, National Academy Press, Washington, D.C. 1990.

Underwood, J. M., 1988, Use of Life-Cycle Costing in the Development of Standards, masters thesis, Naval Postgraduate School, Monterey, Calif., December.

4

GOVERNMENT FACILITIES AND LIFE-CYCLE COSTS

Despite obstacles to the application of life-cycle cost analysis, government agencies have recognized that the analysis process can yield benefits in efficient utilization of resources . . . and they have made progress in achieving control of life-cycle cost. In some cases, notably highway pavement and bridge management, this progress has been substantial and offers lessons that are transferrable to buildings and other facilities management. There are, however, government policies and legislation that discourage effective control of life-cycle costs.

GOVERNMENT POLICIES RECOGNIZING LIFE-CYCLE COST AND ANALYSIS

The economic principles of discounted cash flow analysis that underlie life-cycle cost analysis have been used in certain areas of government decision making for many years, most notably in the development of large dams and other major capital investment projects.[20] Office of Management and Budget (OMB) Circular A-94, issued in 1972, specified the discount rate to be used

[20] The Bureau of Reclamation and the U.S. Army Corps of Engineers were pioneers in the application of engineering economics principles to project evaluation.

in performing these economic analyses and gives limited guidance on how analyses should be conducted.[21]

Sudden increases in petroleum prices and the U.S. energy crisis of the early 1970s were primary factors motivating applications of these principles to buildings. In 1977 President Carter signed Executive Order 12003, "Relating to Energy Policy and Conservation," requiring agencies to prepare plans for cost-effective actions to achieve substantial reductions in energy consumption in existing federal buildings. The 1978 National Energy Conservation Policy Act[22] directed the Federal Energy Administration (FEA) to establish a practical and effective methodology to be used by federal agencies in determining what actions would be effective, on the basis of life-cycle costs and savings. The Department of Energy (DOE), the FEA's successor agency, has continued to foster development and application of life-cycle cost analysis to buildings and has issued mandatory analysis guidelines that apply to more than 400,000 federal buildings (Marshall, 1987).

The DOE has continued to update these guidelines documents and (working with the National Institute of Standards and Technology (NIST)) computer programs intended to assist analysts in performing life-cycle cost studies (DOE, 1990; Lippiatt and Ruegg, 1990). Further amendments of the guidelines, proposed in January 1990 (55 FR 2590) to implement the FEA's Improvement Act of 1988 (P.L. 100–615), are expected to become effective in October 1990. The DOE guidelines are oriented primarily toward actions intended to conserve energy or to enhance use of renewable sources of energy, but the methods described include all nonfuel O&M costs as well as investment costs and salvage values. Higher discount rates are currently applied to nonfuel future costs (in effect giving extra weight to future fuel savings as compared to future operating costs).

Other federal agencies have implemented their own regulations or less formal procedures for life-cycle cost analysis. The General Services Administration, for example, uses a life-cycle analysis in considering whether to lease or purchase buildings for government use and in routine project investment evaluations. The Department of Veterans Affairs (VA) justified the development of its Hospital Building System on the basis of life-cycle cost savings to be achieved. The military construction agencies are frequent users of life-cycle analysis principles in making decisions about facilities intended to support specific mission requirements. While there are no generally applicable or accepted principles and procedures for use of life-cycle cost analysis in controlling the costs of ownership for federal facilities, the guidelines and workshops prepared by DOE, NIST, and the Army's Construction Engineering Research Laboratory have done much to disseminate information on these methods.

---

[21] Circular A-94 (revised), Discount Rates to Be Used in Evaluating Time-Distributed Costs and Benefits, OMB, March 27, 1972.

[22] Title V, Part 3, Sec. 545(a).

State governments use life-cycle cost analysis as well. The state of Maryland, for example, uses these procedures in an effort to achieve "optimal energy use and the lowest possible cost of ownership in State-financed and State-assisted building construction," and the state's Department of General Services maintains guidelines for life-cycle cost accounting.[23] The state of Iowa has enacted a requirement[24] for life-cycle cost analysis as a design criterion in new construction and in renovation of publicly owned facilities "to optimize energy efficiency at an acceptable life-cycle cost." Florida's 1984 Capital Planning and Budgeting Act[25] calls for condition assessments and life-cycle cost evaluations of all state Facilities, updated at least at three-year intervals, as a basis for funding requests to the state's legislature. A 1981 survey identified 26 states that were using life-cycle cost analysis procedures in their facilities programs (Dell'Isola and Kirk, 1981), but, as with federal agencies, there are no generally accepted principles and procedures.

GOVERNMENT POLICIES CONFLICTING WITH LIFE-CYCLE COST CONTROL

Some policies and procedures followed by federal agencies actually work against the goals of achieving low life-cycle costs and controlling the costs of ownership throughout a facility's service life. For example, design fees for federal facilities are limited by law[26] to no more than 6 percent of the estimated cost of construction. This limitation is sometimes used by agency personnel to argue that the development of design alternatives required for life-cycle cost analysis cannot be accomplished within the scope of the designers' contract.[27]

Many agencies require value engineering studies prior to construction, and agency policies typically recognize the potential for life-cycle cost savings

[23] Procedures for Implementation of Life-Cycle Cost Accounting, State of Maryland, Department of General Services, Office of Engineering and Construction, Baltimore, Md., December 1978 (revised May 1980).

[24] Iowa state code, Chapter 470, Life-Cycle Cost Analysis of Public Facilities.

[25] Florida Statutes 1987, Ch. 216, Section 216.015 ff.

[26] This limitation is imposed by five different statutes (Federal Construction Council, 1981).

[27] Committee liaison representatives agreed that this argument is not necessarily valid. Agencies may separately procure services for value engineering, environmental assessment, site investigations, and other activities beyond the scope and budget of the design contract.

at all stages of a facility's service life. However, some agencies routinely include value engineering clauses in their construction contracts, inviting the contractor to conduct his or her own value engineering analysis and to share in any apparent savings realized. Under these value engineering incentive clauses, the contractor's incentives are strongest to reduce first costs, because savings are immediately and obviously realized. Further, funds from construction appropriations may be used to pay contractors for early savings, but these funds are not available to be applied against savings in operations or maintenance.

Legislative budgeting, authorization, and appropriations procedures may also work against controlling life-cycle costs. The lack of distinction in program authorizations between expenditures to construct long-lived facilities and those for procurement of other goods and services with shorter service lives provides strong incentives to reduce a facility's first costs so that other program spending can be maintained. Future expenses for facility O&M may then be raised. Subsequent shortages of funds lead to a deferral of maintenance spending and consequently reduced service life or performance.

Administrative separation of design and construction from operation and maintenance causes similar problems. Staff responsible for administering spending are given incentives to control (usually meaning minimize) their own current expenses, without regard for the concerns of future or prior decision makers.[28] The resulting failure to coordinate action to control life-cycle costs is made more severe by a widespread lack of understanding among designers and maintenance personnel of the concerns the others face. The VA, for example, has found that as much as 20 to 30 percent of the operating efficiencies of their technically sophisticated new hospital and research buildings located on older campuses are lost because current maintenance personnel are untrained and unprepared to deal with the new systems.[29] Training and documentation of design decisions that influence maintenance practice, of some help in overcoming these problems, are hindered by personnel rotation and designers' lack of sensitivity to the demands for maintenance that new systems may present.

[28] Control of life-cycle costs of government facilities constructed with funds raised from bonded indebtedness may be required by the trust indenture or other legal agreements. Sinking funds or other mechanisms may be established to accumulate funds for maintenance and repair, thereby protecting the asset--the building--that secures the bond holders' claim for repayment. Such requirements apply only to revenue bonds or other instruments that tie payments to debt holders to the specific income from the facility or authority.

[29] To deal with this problem, the VA has prepared guidelines for evaluation of existing facilities and a program to develop user manuals for VA projects. These user manuals will help hospital operations staff, administrators, and others to understand the expectations and procedures inherent in design decisions.

Agency officials and designers note that legislative and administrative pressures to limit spending for facilities have reduced levels of performance expected of government facilities, for those aspects of performance not critical to life safety or public health. Standards set for minimum acceptable levels of performance have evolved instead into design targets not to be exceeded. Such standards, applied to the selection of materials, mechanical systems, and other building subsystems that require regular maintenance or replacement with wear, are likely to increase total costs of ownership. Some government agencies, recognizing the risk that future maintenance will be neglected or operating budgets cut, have tended to set higher standards, preferring to increase estimated life-cycle costs somewhat in an effort to avoid uncontrolled future growth of costs of ownership. However, there are no comprehensive studies that document these effects for federal buildings.

IMPROVING THE LIKELIHOOD OF SUCCESS: THE CASE OF HIGHWAY PAVEMENT AND BRIDGE MANAGEMENT

A boom in system expansion in the United States in the 1950s and 1960s that increased dramatically the scale of the nation's investment in highways coincided with rapid advances in computer technology and applications of systems analysis techniques in civil engineering. At the same time, work by development economists at the World Bank and elsewhere began to demonstrate convincingly the direct contribution that pavement conditions have on vehicle operating costs and, in turn, on economic efficiency of a region's transportation system. These forces combined to motivate research and development efforts leading to establishment of practical pavement management systems that, after two decades, are now used routinely by many state transportation agencies to monitor highway facilities, assure maintenance effectiveness, and schedule rehabilitation and replacement of pavements (Hudson, Haas, and Pedigo, 1979).

These management systems, now being extended for use by local government authorities responsible for city and county roadways and for addressing the problems of aging of the nation's highway bridges,[30] offer a potentially useful model for how principles of life-cycle cost analysis can be effectively applied to control costs of ownership of government buildings. Researchers in the field are looking toward evolving present systems into larger integrated "total facilities managements" systems useful to administrators of public facilities responsible for underground services and parks and recreation facilities, as well as road pavements (Haas and Hudson, 1987).

Pavement management systems (i.e., computer programs and data bases) are used to develop and monitor strategies for designing, maintaining, and renewing a highway's pavement to provide at least the minimum acceptable level of road surface performance, subject to the demands of vehicle loads and

[30] See, for example, Hudson et al. (1987).

environmental conditions, at the lowest life-cycle cost compatible with desired reliability and available funds. The management system uses a set of data base and analysis modules that provide information upon which government administrators can base their planning, programming, and budgeting decisions and report on highway performance.

Two results of research and development activities have been key to the successful development of pavement management systems:

- characterization of pavement performance in measurable terms that can be related to economic benefits (e.g., longer pavement life and reduced vehicle operating costs) and
- development of reliable analysis tools that relate performance to characteristics of pavement design, maintenance and repair activities, environmental conditions, and service loads (e.g., vehicle types, weights, and numbers) and which then can be used to predict future performance.

In addition, hardware and software to support computer-based management and analysis of large volumes of data and production of information reports in forms understood by decision makers have been important facilitators of pavement management systems development.

While a highway pavement section is much less complex than a large building, efforts to develop similar management tools for buildings have yielded positive results and benefited from lessons learned in pavement management. These efforts have encountered serious problems because neither performance measurement nor life-cycle analysis models are well developed for buildings and their components. As discussed further in Chapter 5, these are areas that warrant research.

Work done so far seems likely to yield effective results. The U.S. Army Corps of Engineers, for example, has developed a management system for bituminous built-up roofs,[31] and a number of U.S. and international researchers have developed models that would facilitate life-cycle cost management of building energy systems (see Carlsson, 1989, for example). However, these efforts each deal with only a part of the complex multicomponent system that a building represents.

In addition, in sharp distinction with highway pavements, government buildings have no centralized funding mechanism (i.e., analogous to the federally administered fuel tax and construction cost-sharing programs), and no single agency is responsible for construction and management of all facilities. In consequence, it may be more difficult for these steps toward overcoming obstacles to have effective impact on government practices in life-cycle cost management of buildings

[31] The ROOFER system (see Bailey et al., 1989) is being extended to deal with other types of roofs. The American Public Works Association is working with the Corps of Engineers in this effort to bring the system into active use.

REFERENCES

Bailey, D. M., D. E. Brotherson, W. Tobiason, and A. Knehans, 1989, ROOFER: An Engineered Management System (EMS) for Bituminous Built-up Roofs, Report M-90/04, U.S. Army Construction Engineering Research Laboratory, Champaign, Ill. December.

Carlsson, B., 1989, Solar Materials Research and Development: Survey of Service Life Prediction Methods for Materials in Solar Heating and Cooling, Swedish Council for Building Research, Stockholm.

Dell'Isola, A., and S. Kirk, 1981, Life-Cycle Costing for Design Professionals, McGraw-Hill, New York.

Federal Construction Council, 1981, "Review of three recommendations on architect-engineer procurement of the Commission on Government Procurement," Transactions of the Federal Construction Council for 1980–81, National Academy Press, Washington, D.C.

Haas, R., and W. R. Hudson, 1987, Future Prospects for Pavement Management, prepared for Second North American Conference on Managing Pavements, Toronto, November 2–6.

Hudson, S. W., R. F. Carmichael III, L. O. Moser, and W. R. Hudson, 1987, Bridge Management Systems, NCHRP Report 300, Transportation Research Board, Washington, D.C., December.

Hudson, W. R., R. Haas, and R. D. Pedigo, 1979, Pavement Management System Development, NCHRP Report 215, Transportation Research Board, Washington, D.C., November.

Lippiatt, B. C., and R. T. Ruegg, 1990, Energy Prices and Discount Factors for Life-Cycle Cost Analysis 1990, Annual Supplement to NBE Handbook 135 and NBS Special Publication 709, NISTIR 85-3273-4 (Rev. 5/90), U.S. Department of Commerce, National Institute of Standards and Technology, Gaithersburg, Md., May.

Marshall, H. E., 1987, "Building economics in the United States," Construction Management and Economics, vol. 5, pp. S43–S52.

U.S. Department of Energy, 1990, Architect's and Engineer's Guide to Energy Conservation in Existing Buildings, two volumes, DOE/RL/01830PH4, Washington, D.C., April.

5

CONTROLLING LIFE-CYCLE COSTS OF OWNERSHIP

The preceding chapters have discussed a variety of general technical, institutional, and management factors and specific government practices that pose serious obstacles to effective life-cycle cost control. A strong and long-term commitment will be required to overcome these obstacles. This long-term commitment will be built through research and development, and professional education and training and by specific actions that each responsible agency can take.

The committee believes that commitment to life-cycle cost management begins with an understanding that a building's life-cycle is not strictly a series of sequential stages of planning, design, construction, and use. Changes in agency mission, new technology, and random events may define an economic life and lead to substantial new construction or changes in O&M requirements throughout a longer service life. The basic structure of the building may have, in effect, an indefinite service life, while such subsystems as roofs, mechanical equipment, electrical components, plumbing, and interiors may undergo frequent changes and replacements. Life-cycle cost analyses are most frequently used during the planning and preliminary design stages of a project—at the beginning of the project's economic life—but can be used throughout the life-cycle to direct operating and major repair decisions.

The committee's recommendations therefore deal with all aspects of the life-cycle. However, the committee recognizes that the commitment required for effective life-cycle cost management will be difficult to achieve. In the near term, setting guide criteria used in planning and design to limit life-cycle costs may be the only effective tool available to agency professionals.

USING GUIDE CRITERIA[32] TO ACHIEVE LIFE-CYCLE COST CONTROL

The life-cycle costs or future performance of some types of buildings are particularly sensitive to maintenance activities, environmental conditions (e.g., weather), or conditions of use. For example, a leaking roof on a building housing sensitive and costly electronic equipment may have more serious consequences than if the roof were sheltering a bulk storage warehouse. Similarly, neglect of caulking repair and consequent increases in energy use for heating a residential facility located in a temperate climate are likely to be less costly than such neglect would be in a subarctic setting.

Failure to effectively execute the O&M practices implicit in design can sharply increase the costs of ownership or drastically reduce the facility's life-cycle performance. In view of the substantial institutional and management obstacles discussed in Chapters 3 and 4, the committee acknowledges that government agencies may justifiably set their guide criteria to reduce the needs for future maintenance or repair actions to avoid the risk of substantial growth in costs of ownership when these actions are not taken.

However, the committee notes again that setting such criteria is likely to increase the total costs of ownership, compared to levels anticipated with effective life-cycle management. The committee therefore recommends that guide criteria not determined by requirements for life safety, public health, or mission success should be changed on a case-by-case basis, when such change is shown by life-cycle cost analyses to produce savings or to increase certainty that costs can be controlled throughout the service life. An anticipated increase in life-cycle cost to improve the certainty that future costs will be controllable (i.e., that projected life-cycle costs will be realized and not exceeded) will thereby be made a recognized factor in management decision making. The process of choosing to invest more in design and construction to enhance future control is, in turn, consistent with the committee's recommendations for broader action to control the cost of ownership throughout a building's service life.

RECOMMENDATIONS FOR BROADER ACTION

Minimizing life-cycle cost is the most efficient use of resources to achieve facilities needed to support agency missions and the public interest. The benefits of protection and improved productivity of the public's assets warrant the strong measures needed to overcome the obstacles to effective life-cycle cost management.

[32] "Guide criteria" is the term used by many federal agencies to describe their basic facilities requirements. In the private sector, equivalent requirements are typically found in local building codes and manuals of standard practices.

Life-Cycle Cost Management as a Basic Policy

Each agency should formally recognize life-cycle cost management as a basic and essential element of the agency's operating policies. All decisions about facilities must be made within the context of unavoidable uncertainties in what the future agency mission may be and inevitable limitations on capital and operating budgets. The future is unknown and agency missions may change. It may therefore seem impossible to achieve meaningful long-term control of costs of ownership throughout a facility's life-cycle. However, these uncertainties can be explicitly recognized. Applying the principles of life-cycle cost analysis and management with this recognition will improve both the quality of decision making and the efficiency of resource use. (See box.)

Emphasis can be given to this policy by incorporating life-cycle costs and performance into federal design awards programs. Awards for facilities that have delivered 5 to 10 years of appropriate performance at low cost should recognize the entire life-cycle management team--planners, designers, constructors, and O&M personnel. These awards also will establish incentives for data collection to document life-cycle costs.

As guidance to planners and designers who must conduct life-cycle cost analyses and make recommendations to decision makers, agencies should prepare annually updated assessments of the uncertainties and constraints that should be considered in making decisions about facilities. These assessments will be the basis for sensitivity testing or probabilistic methods of characterizing the uncertainty and risk with life-cycle cost analyses[33] and for conclusions based on these analyses.

OPERATIONS, MAINTENANCE, AND CONSTANT IMPROVEMENT

". . . there is a major difference in ways of thinking about machines between us Japanese and Westerners. Westerners tend to think that production equipment is destined to deteriorate as time goes by. For us, though, the installation of a machine is just the beginning, and we strive to improve it so the machine's performance will get better. This improvement is not just the responsibility of engineering experts alone. People on the shop floor know the machine better than anyone else, and so they must cooperate with the engineers. This is the 'suggestion for improvement' system that we employ throughout our global operations."

(Source: Tetsuo sakida, Honda Motor: The Men, the Management, the Machines, Kodansha International, Tokyo, 1982).

[33] See Marshall (1988), for example.

Explicit Design Alternatives

In developing designs and management strategies for facilities, agencies should require their staff and consultants to undertake an explicit analysis of alternatives to explore opportunities for controlling life-cycle costs. Alternatives may be defined as variations on a principal theme or as distinctly different designs or strategies. However, consideration of alternatives is an essential element of life-cycle cost analysis.

The consideration of alternatives is in many cases already performed and documented as part of value engineering or environmental impact studies, but these studies are often performed late in the design process. The committee recommends that the definition of alternatives should be initiated very early in the life-cycle. Documentation of alternatives and reasons for their development or rejection can be a useful resource to the people who will be responsible for subsequent maintenance or retrofitting and general life-cycle cost management.

Designers should be compensated appropriately for the work required to develop well-documented design alternatives. Some agencies may choose to use staff or consultants to perform supplementary studies, outside the scope of the principal design contract, while other agencies will include the life-cycle economic analyses and documentation within the basic design scope. The former approach--life-cycle cost analysis as a separate service--is prevalent in both the private and public sectors. In either case, fees must be adequate to cover the work required and agency budgets should be established that ensure that life-cycle cost analysis is conducted.

Value Engineering for Life-Cycle Cost

Those agencies that use value engineering programs should assure that savings are achieved in expected life-cycle cost rather than construction cost only. The committee has noted the incentives that facility constructors participating in value engineering programs may have to reduce construction cost despite possible increases in future maintenance and repair. Agency staff and their consultants should assure that all value engineering proposals under such programs include a realistic analysis of life-cycle costs and should carefully review these proposals. Training of procurement and professional staff may be needed to facilitate clear understanding of the net impact that value engineering proposals may have on future agency expenses.

The committee proposes that procurement procedures could include bidders' analyses of life-cycle cost as one of the factors for selection in those cases where the supplier may be held accountable throughout a defined economic service life. Warranties or other financial arrangements may be considered to establish incentives for the supplier to control life-cycle cost.[34]

Life-Cycle Management Manuals

Agencies should request that designers prepare a report that can serve as a life-cycle management guidebook for new or substantially renovated facilities. This document--in effect, an owner's manual--would present design decisions made to control life-cycle cost and subsequent operating consequences for each facility. Required future actions needed to assure that the life-cycle proceeds as planned would be explained. Procedures for monitoring condition and performance, as a basis for assuring that these important future actions have been taken and have been effective, would be included as well. The guidebook could be updated as necessary by construction and O&M personnel, throughout the facility's life-cycle, to show the life-cycle cost consequences of new systems and operating conditions.

The guidebook would not necessarily be a traditional notebook or other printed document., rather it might be part of a computer-based information system. Such systems,[35] sometimes termed "integrated data bases," are becoming practical as advances are made in the technologies of computer-aided design and drafting, database management, and artificial intelligence. The life-cycle management manual would be a precursor to complete building management systems, analogous to the pavement and bridge management systems discussed in Chapter 4.

Education and Training

Agencies should encourage exchange and cross-training of design and O&M management personnel to assure that members of each group have practical understanding of life-cycle cost management principles and how life-

[34] Public facilities such as toll roads and water supply systems are already exposed to such incentives by the requirements that revenues cover costs. "Shared savings" programs--still an experimental practice, in which a private firm may invest to make changes in a government facility (e.g., for energy-efficient air conditioning equipment or hospital management) and be repaid from a share of the cost savings realized from those changes--would include such incentives.

[35] The "Woods Hole workshops," a series of meetings sponsored by the Building Research Board from 1983 to 1987, defined many of the principles now being implemented to support integrated data bases. See, for example, Committee on Advanced Technology for Building Design and Engineering (1986).

cycle cost is controlled at all stages of a facility's service life . Personnel from each group should be included in the management team undertaking design, procurement, and subsequent operation and maintenance of each facility. Each cross-trained team should then be held responsible for life-cycle cost control. Continuity in team membership can improve team performance.

The committee proposes that agencies also look beyond these needs for immediate training and seek to encourage professional education for life-cycle cost management. Support to professors in schools of architecture and engineering for research and development of courses and teaching materials on life-cycle cost management could yield substantial long-term benefits to government agencies and the public. Such support could be supplemented by working with professional organizations and accreditation boards to assure that life-cycle cost analysis and management principles are an integral part of the knowledge imparted to building professionals. Professional school accreditation organizations should include life-cycle cost principles as an evaluation factor in accreditation reviews.

Life-Cycle Cost Management Systems

Agencies should work cooperatively to establish life-cycle cost management systems. These systems should be suited to the specific types of facilities that each agency uses, but they should also foster exchange of relevant data to support analysis of life-cycle cost and performance of common subsystems (such as roofs, envelopes, and air conditioning and lighting systems). Development of these systems would draw on lessons learned in pavement and bridge management systems.

Key elements of such systems are an adequate data base on life-cycle performance and cost and reliable tools to enable projection of life-cycle cost and performance of new designs and operating strategies. These systems should produce their information in forms useful for budgeting as well as technical decisions at all stages in a facility's life-cycle. Appendix D presents the committee's initial suggestions of what questions and decisions a building life-cycle cost management system might be designed to address.

The committee recognizes that development of effective and comprehensive life-cycle cost management systems will take time. While a basic common framework for data collection and assembly might be adopted by federal agencies within a 12-to 18-month period, experience with highways indicates that development of an adequate data base would require 5 to 10 years.

While the primary focus of these systems would be on management of individual facilities, early results might be useful in budgeting for management of large portfolios of similar buildings, such as those operated by the U.S. Department of Defense[36] and the General Services Administration. Less data

[36] The U.S. Department of Defense has in fact used this approach in justifying its requests to Congress for maintenance and repair funds.

are needed to predict the percentage of a large inventory likely to require certain actions and the expected costs of failure to take action. These aggregate statistics can be useful in general planning, programming, and budgeting to achieve life-cycle cost control.

High-Level Accountability

Accountability for effective use of maintenance and repair funds and other actions that may invalidate prior decisions on facility life-cycle cost should be assigned at the highest levels in the agency.Because costs for facilities are typically a small part of the total expenses of government programs, effective life-cycle cost control is possible only when senior management are committed to its achievement. The senior management must be willing to forego opportunities either to divert funds to other uses by deferring maintenance or purchasing low-first-cost components or to take advantage of available funds by overinvesting in facilities. In government settings, senior management includes both legislative and executive bodies that may authorize new construction and fail to meet their implied commitment to future costs of ownership.

Agency management must also be willing to present convincingly the principles of life-cycle management to officials who, in the absence of adequate information upon which to base a balanced decision, may be encouraged by popular opinion to make commitments that have lasting consequences for the costs of facilities. Wherever possible, agencies should establish facility endowments, sinking funds, or other formal financial mechanisms to assure adequate resources to implement previous life-cycle management decisions. While economic policy makers often discourage such earmarking of funds, these mechanisms are essential to the management of long-lived facilities in a setting characterized by responses to short-term issues.

CONCLUSION: THE PAYOFFS

The committee recognizes that implementing these recommendations will require significant investments of time, staff effort, and funds; however, substantial payoffs will result.

The committee notes that the aggregate spending on construction of public buildings in 1989 exceeded $30 billion (U.S. Department of Commerce, 1990). Using common rules of thumb, the committee estimates that equivalent annual capital cost[37] for this new construction would be about $400 million.

[37] This is the equivalent annual spending which, over the facilities' service lives, has approximately the same present value as the construction cost. This is conceptually equivalent to the mortgage payments a homeowner might make.

Expenses for maintenance and repair might account for an additional $300 million to $600 million annually. Total annual spending for management of public buildings may then be about $0.7 billion to $1 billion.[38]

If development and application of life-cycle cost management principles can produce savings of only one-half of 1 percent of these amounts, the public's annual return on investment in research and development could easily exceed $3.5 million. Increases in productivity of activities housed in better-managed facilities could be even greater. The committee is therefore confident that the costs of implementing its recommendations will be repaid many times over in savings in the nation's costs of public facilities.

REFERENCES

Committee on Advanced Technology for Building Design and Engineering, 1986, Report from the 1985 Workshop on Advanced Technology for Building Design and Engineering, Building Research Board, National Academy Press, Washington, D.C.

Marshall, H. E., 1988, Techniques for Treating Uncertainty and Risk in Economic Evaluation of Building Investments, NIST Special Publication 757, U.S. Department of Commerce, U.S. Government Printing Office, Washington, D.C., September.

U.S. Department of Commerce, 1990, Construction Review, vol. 36, no. 2, March/April.

[38] This analysis reflects an implicit assumption that all public construction is undertaken to replace existing facilities (i.e., that there is no net expansion in the inventory of public buildings. If the inventory is not being effectively replaced or if the average age of facilities is changing significantly, the estimates may be low.

APPENDIX A

GLOSSARY

The following definitions are specific to life-cycle cost analysis and building economics, as discussed by the committee. Readers may wish to refer to the several texts and articles cited throughout this report for additional information or detailed discussion of these terms.

BENEFITS. The shelter, services, and other mission support a building or other facility provides to its owners and users throughout the facility's *lifetime*; generally assumed to be positive and may be valued in monetary or other terms.

BUILDING PORTFOLIO. A collection of buildings or other constructed facilities managed by a single agency or other owner.

COSTS. Expenditures of funds required at a particular time to obtain the *benefits* of a facility, valued in monetary terms; not the same as *disbenefits*.

COSTS OF OWNERSHIP. The total of all costs incurred, generally by the owners but also by the users, to obtain the benefits of a facility.

DESIGNERS. Architects, engineers, and other professionals responsible for making technical recommendations about a facility's configuration, materials, mechanical systems, and other characteristics that determine future performance and cost.

DISBENEFITS. Undesirable results obtained by a building's owners or users as a result of the building's *performance*; negative *benefits*; not the same as **costs**, although the terms are sometimes used interchangeably.

DISCOUNT RATE. A measure of the economic time value of money, the opportunity cost of having funds or benefits available now versus at some future date; may include allowance for *inflation* (nominal discount rate); generally expressed as an annual percentage rate; not the same as *interest rate*.

ECONOMIC COSTS. Costs expressed in terms of their economic value, such that comparisons can be made among costs incurred in the past, present, or future.

ECONOMIC LIFE. The period of time over which *costs* are incurred and *benefits* or *disbenefits* are delivered to an owner; an assumed value sometimes established by tax regulations or other legal requirements or accounting standards and not necessarily related to the likely *service* life of a facility or subsystems.

EQUIVALENT ANNUAL VALUE. The amount of *costs* or *benefits* that, if received as a uniform annual amount for the duration of the *economic life*, would have a present value equal to all costs or benefits anticipated; computed using the *discount rate*; see *economic cost, discount rate*, and *present value*.

FINANCIAL COSTS. *Costs* expressed in terms of monetary value at the time incurred.

INFLATION. A rise in the general price level; also described as a general decrease in purchasing power of a given amount of funds.

INTEREST RATE. The percentage cost incurred for use of funds (typically borrowed) over some period of time; typically expressed on an annual basis; typically includes amounts to account for expected *inflation*, time value of money (see real *discount rate*), and compensation for risk and administrative effort of lender who provides funds.

LIFE-CYCLE. The sequence of events in planning, design, construction, use, and disposal (e.g., through sale, demolition, substantial renovation) during the *economic* or *service life* of a facility; may include changes in use and reconstruction.

LIFE-CYCLE COSTS. The *present value* of all anticipated *costs* to be incurred during a facility's *economic life*.

NET PRESENT VALUE. The sum of the *present values* of all *costs* and monetary-valued *benefits* of a facility over its *economic life*.

NONMONETARY COSTS. Disbenefits not readily measurable in monetary terms, such as air pollution emissions or worker absenteeism.

NONRECURRING COSTS. *Costs* incurred once, infrequently, or on an irregular basis during a facility's economic life, typically for repair or replacement of components or subsystems.

NONQUANTIFIABLE COSTS. *Disbenefits* not readily measurable but attributed to a facility's performance, such as worker dissatisfaction or loss of readiness.

PERFORMANCE. The degree to which a building or other facility serves its users and fulfills the purpose for which it was built or acquired.

PORTFOLIO. See *building portfolio.*

PRESENT VALUE. The amount of *costs* or *benefits* that would be equivalent, if incurred or received now, to an amount incurred or received in the past

or anticipated in the future; computed using the *discount rate*; see *economic cost.*

RECURRING COSTS. *Costs* incurred on a recurring and generally regular basis throughout a facility's economic life, typically for operation, normal maintenance, and anticipated repair or replacement of building components or subsystems.

SERVICE LIFE. The period of time over which a building, component, or subsystem provides adequate performance; a technical parameter that depends on design, construction quality, operations and maintenance practices, use, and environmental factors; not the same as *economic life.*

TIME HORIZON. The period of time considered by an analyst or decision maker in choosing among alternative designs or management strategies; typically the *economic life* for such facilities decisions as lease or purchase, build or buy, and renew or replace.

VALUE ENGINEERING. Currently defined by most federal agencies as an organized effort directed at analyzing the function of construction operations, systems, equipment, facilities, procedures, methods, and supplies for the cost consistent with the requirements for performance, reliability, quality, safety, and maintainability.

APPENDIX B

BIOGRAPHICAL SKETCHES OF COMMITTEE MEMBERS AND STAFF

EDWARD COHEN (Chairman), is managing partner of Ammann and Whitney, Consulting Engineers, and a member of the National Academy of Engineering. He earned a B.S. in engineering from Columbia University in 1945 and an M.S. in civil engineering in 1954. An internationally recognized structural consulting engineer, he has received many awards for his achievements in engineering, including the Egleston Medal from Columbia University, 1981; Goethals Medal for Engineering Achievement from the Society of American Military Engineers, 1985; and the Academy Award from the New York Academy of Sciences, 1989. He has been named to honorary membership of the American Society of Civil Engineers, the American Concrete Institute, and the New York Academy of Sciences. His company was responsible for the engineering of the Verrazano Narrows Bridge, Washington's Dulles International Airport, and restoration of the Statue of Liberty and the West Face of the U.S. Capitol.

GREGORY B. COLEMAN is vice president of the American Consulting Engineers Council Research and Management Foundation. He received a B.S. in engineering from the University of Florida in 1975. His work experience includes private consulting and project management for major projects at home and abroad. He is currently responsible for developing, managing, and carrying out research and educational projects related to engineering management, design criteria for facilities, building energy systems and analysis, building operations and maintenance, and microcomputer hardware and software applications for engineers.

NORMAN G. DELBRIDGE, JR., PE, received a B.S. degree from the U.S. Military Academy in 1953 and an M.S. in civil engineering from Iowa State University in 1957. As a career Corps of Engineers officer, Mr. Delbridge was responsible for management of major construction programs at home and abroad, retiring as deputy chief of engineers in 1986. He serves as a consultant to companies involved in construction services and educational institutions and on national policy groups, including Transportation Research Board committees.

W. RONALD HUDSON is the Dewitt C. Greer Centennial Professor at the University of Texas at Austin. He holds a Ph.D. in civil engineering and is a registered professional engineer. He is an internationally recognized expert in highway and transportation facilities, particularly in the application of management and life-cycle costs concepts to pavements and bridges, and has 24 years of research and teaching experience and 10 years of design and field experience. Through participation in many professional and technical committee activities, he has contributed extensively to his profession and has been awarded many honors, including the Highway Research Board Award. He has published over 250 technical reports and papers and has given more than 300 oral presentations and invited lectures at national and international meetings. His publications include the book Pavement Management Systems, published in English and Japanese, and over 50 articles related to life-cycle and system analysis.

ROBERT E. JOHNSON is chair of the doctoral program in architecture and associate professor of architecture at the College of Architecture and Urban Planning of the University of Michigan. He received an A.B. in economics from Colgate University in 1968, a B.Arch in 1973 and M.Arch. in 1974 from Syracuse University, and D.Arch in 1977 from the University of Michigan. Building economics is one of Dr. Johnson's teaching and research areas. His book The Economics of Building was recently published by John Wiley & Sons, Inc. He has served on the Building Research Board's Committee on Budget Estimating Techniques.

ALEXANDER I. OUMOV is a multilingual executive architect with over 25 years of experience in design, management, and profitability of largescale projects in the domestic and international fields. Formerly director of architecture, design, and construction with the Hotel Services Division of Holiday Corporation, he received his bachelor of architecture from Alexandria University and attended the Ecole des Beaux Arts in Paris.

JAMES G. PALMBORG is with Sverdrup Corporation, working on operations and maintenance of new U.S. embassy buildings. He earned a B.Sc. in mechanical engineering from Tufts University and a master of science in financial management from NPS Monterey. He has 16 years of experience in facilities management/public works, including planning, engineering, construction. maintenance and repair, and transportation. He also has comprehensive experience in the management and operation of buildings, grounds, and utility systems. Mr. Palmborg served as a liaison representative on the Building Research Board's Committee on Advanced Maintenance Concepts for Buildings.

GARY L. REYNOLDS is director of facilities management and a former temporary assistant professor of mechanical engineering at Iowa State University. He is currently a faculty member of the Association of Physical Plant Administrators Facilities Management Institute. He received a B.S. in engineering science in 1972 and an M.S. in mechanical engineering in 1979 from Iowa State University. Mr. Reynolds is coauthor of the manual used for the Iowa Class A Energy Auditor Training program, which was subsequently used as the basis for the state of Iowa's life-cycle costing legislation. His current position involves oversight of the application of the life-cycle costing law to buildings constructed for Iowa State University.

VICTOR E. SANVIDO is assistant professor of architectural engineering at Penn State. He was awarded a B.Sc. in civil engineering from the University of Cape Town in 1980, an M.S. in 1982, and a Ph.D. in 1984 in civil engineering/ construction engineering and management from Stanford University. Dr. Sanvido has a great deal of field experience in the construction industry and, as a scholar and educator, has focused on productivity improvement and construction automation through modeling the construction project life-cycle process and developing tools for its integration. He has successfully applied these tools to construction projects.

GAYLAND B. WITHERSPOON is associate dean of the College of Architecture, Clemson University, and is a principal of a limited part-time professional practice. He received a bachelor of architecture degree from the University of Arkansas in 1956 and a master of science in architecture with building structure's theory option from the University of Illinois in 1961. During 9 years of active duty and 23 years of reserve duty with the U.S. Air Force he gained considerable experience in facilities' planning, design, and construction. He is actively involved in professional societies at the local, state, and national levels.

JAMES E. WOODS is the William E. Jamerson Professor of Building Construction at the College of Architecture and Urban Studies at Virginia Polytechnic Institute and State University. Formerly, he was senior staff scientist at Honeywell Physical Science Center. He received a bachelor of science degree in mechanical engineering from the University of New Mexico in 1962 and was awarded a master of science degree in physiological sciences in 1971 and a doctorate in mechanical engineering in 1974 from Kansas State University. Dr. Woods is responsible for research and development in indoor air quality. He has lectured and written extensively on indoor air quality and its effects on health, comfort, and energy consumption as well as its cost implications and the role of designers, who must balance the sometimes conflicting environmental and economic objectives of both its occupants and its owners.

STAFF

ANDREW C. LEMER, director, is an engineer-economist and planner. Formerly division vice president with PRC Engineering, Inc., Dr. Lemer is founder and president of the MATRIX Group, Inc., and has written widely on matters of building economics and development policy, often in conjunction with his work on major projects in the United States and overseas. He received his S.B., S.M., and Ph.D. degrees in civil engineering from the Massachusetts Institute of Technology and is a member of the American Institute of Certified Planners, the American Society of Civil Engineers, the Urban Land Institute, and the American Macroengineering Society.

PETER H. SMEALLIE, senior program officer, and executive secretary of the Public Facilities Council has a B.A. in urban studies from St. Lawrence University. He has served as vice president of Thomas Vonier Associates, an architecture and consulting firm, and was a program director with the American Institute of Architects Research Corporation. He recently completed a book titled New Construction for Older Buildings: A Design Sourcebook for Architects for the publisher John Wiley & Sons.

APPENDIX C

OVERVIEW OF LIFE-CYCLE COST ANALYSIS PROCEDURE

This brief presentation of the principles of life-cycle cost analysis is intended as background for the committee's report. Readers seeking a complete discussion of the topic should refer to the literature. Several of the texts available are cited as references in this report. Readers also should refer to Appendix B (Glossary) of this report.

THE BASIC PREMISE

Life-cycle cost analysis is based on the economic principles of discounted cash flow, which allow the analyst to express the value of money spent at any particular time in terms of an equivalent amount spent at any other time. Using these principles, one may compare the costs of purchasing (for example) an electric battery that will last for 4 years versus purchasing one that will last only 1 year and replacing it three times to obtain 4 full years of service. The principles of discounted cash flow tell the decision maker that unless the cost of the 4-year battery is less than four times the cost of a 1-year battery.[39] the 1-year choice is probably better because the total economic costs are lower for the same performance.

[39] All other things being equal and neglecting costs for installation and time out of service, the future costs are simply "discounted," using the discount rate, to their equivalent present value. The preferred choice is the one with the lowest present value of costs.

Life-cycle cost analysis is a process of estimating all of the costs likely to be incurred over the economic life of a facility and expressing those costs in terms of an equivalent single value that facilitates comparisons among alternative designs, operating strategies, and external conditions that may influence costs. Life-cycle cost analyses of buildings generally are made to consider whether increased expenditures for construction (e.g., for more durable materials or equipment that uses less energy) are warranted by likely savings in future operating or maintenance costs. Guidelines have been developed for life-cycle cost analysis (ASTM, 1988, for example), but there are no universally accepted procedures that assure that all analyses reflect similar definitions or assumptions.

As with most predictive methods, life-cycle cost analysis requires assumptions about the economic life (also termed the time horizon) and the discount rate.[40] A longer time horizon will lower the annual savings required to justify an increased initial capital investment and, consequently, the shorter the life of a building, the less worthwhile it is to invest in actions to reduce future operating or maintenance costs. Higher discount rates require larger annual savings to justify an initial capital investment and, consequently, the higher the discount rate, the less worthwhile it is to invest in initial capital costs in order to reduce future operating and maintenance costs.

Life-cycle cost analysis specifically requires the consideration of all significant costs of purchasing, owning, operating, and disposing of a facility. Savings are measured as negative costs, and analysts may sometimes include benefits or disbenefits not directly related to the building.

Costs are generally classified as recurring or nonrecurring. This classification is useful for applying the appropriate mathematical equations used to compute equivalent values in performing the economic analysis. Funds are typically assumed to be spent at the end of the year in which costs are incurred. However, a comprehensive analysis generally includes all of the following elements: (1) initial capital cost, (2) annual operating cost (energy and maintenance), (3) periodic replacements, (4) additions and alterations, (5) use costs, and (6) salvage value. The single-number result of life-cycle cost analysis is typically either the present value or equivalent annual value of all costs anticipated for the economic life.

UNDERLYING ASSUMPTIONS

Life-cycle cost analysis is typically applied to a number of alternative plausible designs or management strategies that are generally equivalent except for their different patterns of cash flow during their economic life. The intent of the analysis is to find an alternative that balances initial procurement costs

---

[40] The distinction between interest rate and discount rate is important in economic analyses of this sort. See Appendix B.

and later operations costs to provide satisfactory performance at the lowest life-cycle cost. This analysis depends on several important assumptions.

First, there must be plausible alternatives or variations on the design or strategy being considered. These alternatives have to be comparable and more or less equally practical.

More specifically, noneconomic performance factors are assumed to be equal or irrelevant for all of the design alternatives. There is no explicit mechanism for considering intangibles that may have a bearing on which alternative is preferred, and the analysis most typically includes only financial costs. All costs (and benefits, if they are included) must be measurable as monetary values to be included in the life-cycle cost computation.

Only significant differential costs need to be considered in distinguishing among alternatives. If the alternatives were, for example, building designs being proposed for the same site, the cost of the site would not be included in the analysis.

The analysis assumes that there is a free trade-off between initial capital costs and long-term costs. That is, the long-term annual savings that will result from the decision to spend more on a more durable or easier to maintain building can more than offset the extra initial capital cost of that building.

Because inflation is common to all alternatives it is usually not considered a differential cost. The life-cycle cost analysis is generally performed using the assumption of constant dollars. As with other economic evaluation methods, the availability and reliability of cost data are significant practical limitations on the value of the analysis results.

THE ANALYSIS PROCESS

Most life-cycle cost analyses are conducted within the context of the traditional design or problem-solving process: (1) define objectives, (2) identify alternatives, (3) define assumptions, (4) project benefits and costs, (5) evaluate alternatives, and (6) decide among alternatives. The estimated life-cycle cost of each alternative is one of the several factors typically considered in evaluating and deciding among alternatives.

Many building-related life-cycle cost studies focus on improving the performance of specific building components likely to be most susceptible to trade-offs between construction and operation and maintenance activities, for example, HVAC (heating, ventilating, and air conditioning) systems. Defining objectives includes defining such a focus (which may be the building as a whole), identifying the people in an organization who will be affected by the study (and who should therefore be involved), and developing clear and measurable criteria for judging the effectiveness of suggested alternatives. It is important to state the study objectives in such a way that they do not limit solutions.

While the focus of a life-cycle cost analysis may be on the cost implications of various technical design decisions, it is important to consider the impact of building design decisions on the organization. In some situations potential problem solutions may require more than the replacement of a building component, and it will be important to identify those individuals who would be affected and include their perspectives in the analysis.

Life-cycle cost analysis focuses on evaluating and comparing alternatives, but it provides no guidance as to how alternatives should be generated. The ability to develop appropriate alternatives is a function of the creativity of the design and management team. However, two criteria can guide the search for alternatives within a life-cycle cost context: (1) alternatives should represent a range of possible solutions to the identified problem; interdisciplinary teams are often helpful in meeting this criterion because of the members' different backgrounds and experiences; and (2) because life-cycle analysis is limited to financial and economic factors, the alternatives being compared should have approximately the same characteristics of performance.

The alternative with the lowest total life-cycle cost is typically preferred. However, other criteria such as risk minimization, corporate image or public role of the building, ease of implementation, and other judgmental factors can become a significant part of the selection process.

Because of the uncertainty associated with estimating the future, sensitivity analysis of life-cycle cost estimates is essential to test the effects of changes in such critical parameters as the discount rate and timing of major operation or repair expenditures. Sensitivity analysis is typically undertaken at the end of the analysis process, to confirm that the conclusions would remain valid if different, but still reasonable, assumptions had been used.

APPLICATIONS

Life-cycle cost analysis has typically been used to evaluate building design alternatives from a particular and limited technical perspective. For example, energy for heating, air conditioning, and lighting accounts for a large share of operating costs and has therefore been the focus of research on strategies for minimizing life-cycle energy costs through investments in energy-saving equipment.

However, life-cycle cost analysis could have much broader application within a context of business productivity. Facilities-related costs are typically a small fraction of the costs associated with economic and social activities that buildings house, but the influence of a building's performance on these activities can be substantial. If the ventilation system does not work effectively, for example, a building's occupants may suffer health problems, resulting in declining production, lost work time, and health care expenses. Failure of a roof may lead to costly damage to the contents of the building. Life-cycle cost analysis may also be used to assess the impact of the building's design on an enterprise, taking into account such costs as staff salaries, lost construction time, fire insurance, and lost revenues due to down time. Some researchers have explored these relationships, and some major industrial corporations consider them in making decisions about facilities, but data and analysis tools are not yet available to support this broader application of life-cycle cost analysis.

REFERENCE

ASTM (American Society for Testing and Materials), 1988, Standard Guide for Selecting Economic Methods for Evaluating Investments in Buildings and Building Systems , E 1185–87, ASTM, Philadelphia, PA., January.

APPENDIX D

ISSUES ADDRESSED BY A BUILDING LIFE-CYCLE COST MANAGEMENT SYSTEM

A building life-cycle cost management system could be useful in addressing a number of issues raised in decision making at legislative, administrative, and technical levels. The following list represents the committee's suggestions of these issues that a life-cycle cost management system might be designed to address.

LEGISLATIVE LEVEL

1. **Justification of budget requests**. Legislators are faced with a variety of competing demands and those that "make the case" in a clear, properly supported manner are likely to receive more favorable consideration.
2. **Effects of reduced capital or maintenance funding**. Legislators may ask what the short-and long-term effects are of less funds, perhaps even zero capital budget, and should be able to get answers to questions related to deterioration of service, extra maintenance, early replacement costs, and effects on users.
3. **Effects of deferring work or lowering service standards**. The frequent consequence of reduced funding is deferral of maintenance and rehabilitation, or lowering the standards of acceptable service; a management system should forecast the effects of such decisions.
4. **Effects of budget request on future status of the portfolio**. If a program funding request is not approved, will the average building service conditions deteriorate? Alternatively, what level of funding is required to keep the portfolio in its present condition or to reduce a backlog of deferred maintenance?

5. **Effects of increased utilization and occupancy**. How will service conditions and operating and maintenance costs be affected?

ADMINISTRATIVE LEVEL

1. **Priority programming of improvements**. Enhanced justification for capital and operating budget requests.
2. **Summary condition assessment of the current portfolio**. Objective aggregate measures, with graphical and tabular displays, based on inventory measurements.
3. **Response to legislative inquiries**. Quantitative estimation of effects of alternate budget levels.
4. **Management planning**. Quantitative assessment of effects of deferring maintenance or rehabilitation, future status of portfolio under alternate funding scenarios, and manpower requirements.
5. **Interface of facility management and overall agency management**. Budget coordination and justification, procurement planning, and staffing logistics.

TECHNICAL LEVEL

1. **Portfolio condition and operations performance data**. On-line monitoring of operations and maintenance activities and facility condition.
2. **Projected maintenance planning**. Forecasting response to use and occupancy, condition survey, routine practices, and external factors (e.g., energy prices).
3. **Service standards**. Assessing criteria for minimum service, maximum loading or use, and maximum distress.
4. **Purchasing and retrofit decisions**. Consequences of new materials and products, bulk purchasing, and replacement by retrofit versus attrition.